ALL ABOUT SHARKS

WEIRD SHARKS

by Chelsea Xie

BrightPoint Press

San Diego, CA

© 2023 BrightPoint Press
an imprint of ReferencePoint Press, Inc.
Printed in the United States

For more information, contact:
BrightPoint Press
PO Box 27779
San Diego, CA 92198
www.BrightPointPress.com

LIBRARY OF CONGRESS CATALOGING-IN-PUBLICATION DATA

Names: Xie, Chelsea, author.
Title: Weird sharks / by Chelsea Xie.
Description: San Diego, CA : BrightPoint Press, [2023] | Series: All about sharks | Includes bibliographical references and index. | Audience: Grades 10-12
Identifiers: LCCN 2022003798 (print) | LCCN 2022003799 (eBook) | ISBN 9781678203726 (hardcover) | ISBN 9781678203733 (eBook)
Subjects: LCSH: Sharks--Juvenile literature. | Sharks--Anatomy--Juvenile literature.
Classification: LCC QL638.9 .X544 2023 (print) | LCC QL638.9 (eBook) | DDC 597.3--dc23/eng/20220127
LC record available at https://lccn.loc.gov/2022003798
LC eBook record available at https://lccn.loc.gov/2022003799

CONTENTS

- Hammerhead sharks have heads that are shaped like hammers. Their head shape helps these sharks quickly swim to the surface or dive. It also gives them better eyesight than other sharks.

- Goblin sharks have quick jaws. They can move their lower jaws like slingshots to latch onto prey.

- Thresher sharks use their long tails to herd prey close together. They whip their tails at prey to stun or even kill them.

- Some sharks are filter feeders. They swallow large amounts of water. Gill rakers separate krill and plankton from the water. Then the water is pushed out through their gills.

- Sharks may have unusual diets. Tiger sharks will take bites out of trash. Bonnethead sharks eat seagrasses.

- Deep-sea sharks live in cold, dark waters. The Greenland shark has a chemical in its blood that keeps it from freezing. Some deep-sea sharks glow to attract prey.

- Sharks like the wobbegong and angel shark hide on the ocean floor. They wait for prey to swim close. Then they attack when the prey is within range.

- Some sharks use their fins to walk on the ocean floor. The epaulette shark is even able to use its fins to walk on land for short periods of time.

WHIP-TAILED SHARK

A common thresher shark prowls through the warm waters of the Pacific Ocean. Its massive tail swings behind it. This tail is one of the most recognizable of all shark tails. As with many sharks, the tail is split into two lobes with pointed ends. But the extreme length of the thresher shark's tail

sets it apart. The top lobe extends for nearly
10 feet (3 m). It makes up more than half of
the shark's length.

The thresher shark uses its tail to swim
through the ocean. But it also serves
as a deadly weapon. The shark senses
movement in the water. With a sweep of
its tail, the thresher shark hurtles toward a
school of sardines. It herds the fish close to

Thresher sharks can grow to be 20 feet (6 m) long.

the surface. Then it whips its tail at the fish, stunning them. The sardines can no longer escape. In a burst of speed, the thresher shark charges at the fish from below. Strong jaws with many rows of teeth crunch down on the meal.

WEIRDER AND WEIRDER

The thresher shark with its long tail is just one of the many strange sharks that swim through Earth's oceans. Other shark **species** have unique head shapes. Sharks in the deep ocean may have even weirder appearances. They may look eel-like. They may glow in the dark water. Sharks that

Thresher sharks often hunt schools of fish, such as sardines. They will also eat squid, crustaceans, and seabirds.

spend their lives at reef bottoms may be flat and pancake shaped.

Many sharks eat animals such as fish and squid. Others have more unusual diets that include tiny creatures called plankton. Some sharks even eat plants. Sharks may look and act strangely. But these features and behaviors help them thrive in their habitats.

1
SHARK HEADS AND SHARK TAILS

Many sharks have torpedo-shaped bodies. Pointed heads allow them to swim easily through the water. But sharks come in many shapes. They may have unusual heads, jaws, and tails. These **adaptations** help sharks hunt and survive.

HAMMERHEAD SHARKS

There are ten species of hammerhead sharks. They vary in size. The smallest is the scalloped bonnethead. It grows to be just 3 feet (1 m) long. The great hammerhead is

There are ten different species of hammerhead shark.

the largest hammerhead species, reaching

20 feet (6 m) in length.

Hammerhead sharks are named for

their wide heads, which are shaped like

hammers. The winghead shark has the

largest head of all hammerheads. Its head is

3.3 feet (1 m) wide, which is half of its body

length. The head of a bonnethead shark

is much smaller. Its head is narrow and

shaped like a shovel.

The broad, flat heads of hammerheads

create a lot of drag. Drag is a force caused

by water pushing against something moving

through it. This can make it difficult to swim

quickly. Because of its head shape, some

hammerhead sharks experience ten times

more drag than other types of sharks.

But this unique head gives the shark

other advantages while swimming. Glenn R.

Parsons studies ocean life. He has studied

Some hammerheads use their wide heads to ram prey.

the hammerhead shark's head shape.

He found that the shape helps the shark

maneuver. "They make these real quick,

kind of jerky turns," Parsons described.[1]

When a hammerhead raises or lowers its

head, another force comes into play. This force is called lift. Water presses on the shark's head. This pressure allows the hammerhead to swim to the surface or dive quickly.

The shape of a hammerhead shark's head improves the shark's vision. Most sharks have eyes at the front of their heads. Their eyes are relatively close together. The eyes of a hammerhead shark are set far apart. One is located on each side of the hammerhead. This gives the shark a wide visual field. A visual field is the total area an animal can see at once. Having a wide visual field means a hammerhead can see to its sides without turning its head.

Hammerhead sharks also have good depth perception. They can tell how close or far away an object is. Like other sharks, hammerheads have binocular vision. This means each eye has its own visual field. When the visual field of each eye overlaps, it creates a sense of depth. This overlap is greater in hammerhead sharks.

The hammer-shaped head does not only improve the shark's vision. It also helps a sense called electroreception. This is a sense that all sharks have. It allows sharks

SHARK VISION

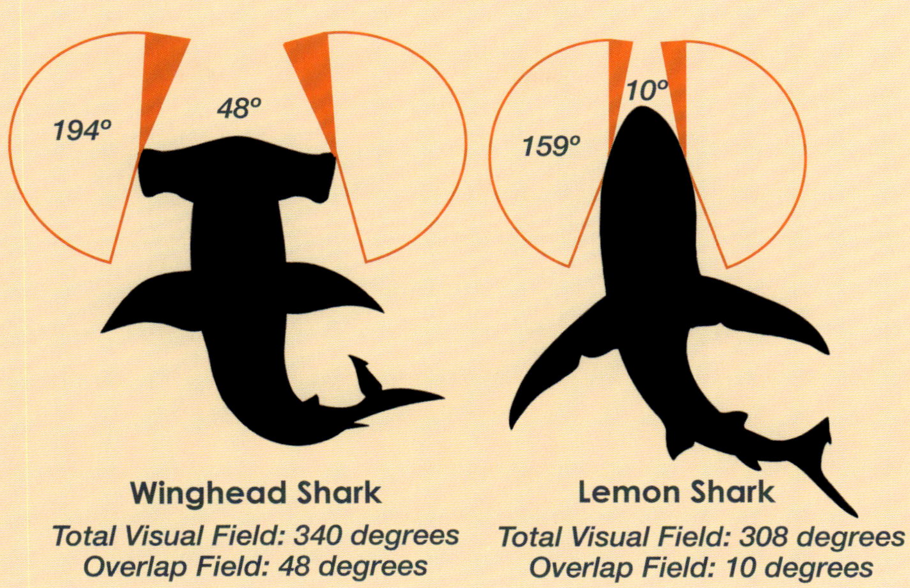

A winghead shark's eyes are set far apart. It has a wider visual field and better depth perception than a lemon shark. A lemon shark has a more typical head shape for a shark.

to detect electrical fields in the water.

Animals give off faint electrical pulses when

they move. Their heartbeats also give off

electrical pulses. Electroreception allows

sharks to find **prey**. Sharks use sensory

organs called ampullae of Lorenzini for electroreception. These organs are found throughout a hammerhead's head. They allow the shark to easily locate prey such as stingrays that may hide under the sand.

GOBLIN SHARKS

Goblin sharks live mostly in the deep ocean. They only come near the surface at night. Because of this, they are rarely seen by humans. Scientists are still learning more about these creatures.

Goblin sharks might be some of the strangest-looking sharks in the ocean.

Most goblin sharks are 10 to 12 feet (3 to 3.7 m) long. However, scientists believe they might be able to grow even larger.

They have long snouts and flabby

bodies. Their stubby fins make them slow

swimmers. Their jaws stick out from their

heads and are filled with crooked teeth.

Goblin sharks may be slow-moving,

but they are fearsome **predators**. They

have the fastest jaws of any fish. Goblin

sharks can move their jaws at 10 feet per

PINK SHARKS

Goblin sharks have skin that is partially see-through. They look pink because their blood is visible under their skin. Scientists study goblin sharks that are accidentally killed in fishing nets. The sharks they study are gray or brown because the blood has dried out.

second (3 m/s). Their lower jaws move like a slingshot. Prey becomes trapped in their claw-like teeth. This fast movement surprises the animals they feed on, which include squid and fish.

THRESHER SHARKS

While goblin sharks use their quick jaws for hunting, other sharks depend on their powerful tails. There are three types of thresher shark. Each species has a tail that is as long as the rest of its body. Thresher sharks feed mainly on fish that gather in groups called schools.

A thresher shark uses its long tail to herd fish close together. Then it whips its tail at its prey. Its tail speed can reach 50 miles per hour (80 kmh). It can stun and even kill fish with a crack of its tail.

These powerful sharks are also known to breach, or jump out of the water. Thresher sharks sometimes eat seabirds. The sharks attack and kill the birds with their tails.

Common thresher sharks often swim near the surface. But they sometimes dive as deep as 1,800 feet (550 m) below the surface. The deep ocean is very cold. Most sharks are cold-blooded. This means that their body temperatures match the temperature of their surroundings. They cannot survive if they get too cold. Thresher sharks have an adaptation that lets them survive in the cold waters. Unlike most sharks, thresher sharks can somewhat control their body temperature. They quickly shiver, which keeps them warm.

WHY SHARKS MIGRATE

Common thresher sharks are found worldwide. They **migrate** through the oceans. Sharks migrate for many reasons. Some travel to warmer waters to give birth. Others migrate to follow food sources. Sharks need to keep their bodies at a constant temperature. Because they are cold-blooded, sharks migrate as water temperatures change.

2
FILTER FEEDERS AND WEIRD EATERS

Sharks are known for being fierce predators. Some sharks hunt large animals such as seals. Others eat smaller prey like fish and squid. Filter-feeding sharks eat animals too. But instead of eating large fish, they feed on small fish, shrimp, and tiny creatures called plankton. The ocean is home to sharks with unusual appetites.

WHALE SHARKS

Whale sharks have thick, blue-gray skin.

They have white spots on their backs. Each whale shark has a unique pattern of spots.

Sunlight reflects off the spots. This helps whale sharks blend in with the ocean.

Whale sharks are the largest fish on Earth. However, they are not dangerous to humans.

Whale sharks are the largest shark species in the world. They can grow to be longer than 40 feet (12 m). Despite their large size, they eat some of the smallest creatures in the ocean. They eat plankton and krill. Whale sharks are filter feeders.

Filter feeders often swim with their mouths open. They suck in large amounts

of water. Gill rakers in their throats trap the
plankton and krill. Then, the water is pushed
out through their gills. The filter feeder eats
the remaining prey. Whale sharks filter more
than 1,585 gallons (6,000 L) in an hour.

Whale sharks typically travel at just
3 miles per hour (5 kmh). Remora fish may
travel near these sharks. Remoras attach
themselves to the whale shark's skin. They
eat **parasites** off of large animals, including
whale sharks. In turn, whale sharks offer
remoras protection. Predators do not
usually bother adult whale sharks because
of their size.

Because whale sharks only eat tiny ocean creatures, they are not a threat to people. Divers sometimes swim with these gentle giants. Brad Norman is an ocean scientist. He has been studying whale sharks for over twenty-five years. He recalls swimming with whale sharks. "It was one of the most amazing experiences I've ever had," he remembered. "I'll never forget it."[2]

BASKING SHARKS

Basking sharks are another large species of shark. They look similar to great white sharks. But basking sharks are harmless filter feeders. They are nicknamed sun fish because they spend most of their time near the surface. Scientists have even seen basking sharks breach the surface. The sharks may leap out of the water to remove parasites from their bodies.

These filter feeders are known for their huge mouths. Their mouths are 3 feet (1 m) wide and often hang open to eat plankton. Their gill rakers separate prey from water and are replaced each year.

MEGAMOUTH SHARKS

The megamouth shark is the smallest filter-feeding shark. It grows to just 16 feet (5 m) long. But as its name suggests, it has an impressive mouth. Its mouth can stretch as wide as 4.25 feet (1.3 m). Megamouth sharks can swallow 150 gallons (680 L) of water in a single gulp.

Megamouth sharks are a rare species. Scientists didn't discover this shark until 1976. By 2021, researchers had studied fewer than sixty of them. This is partly because the megamouth shark lives deep underwater. It can survive at depths of 15,000 feet (4,600 m). It swims closer to the surface at night.

SIXGILL SHARKS

Sixgill sharks live in the deep ocean. Little is known about these sharks. In a video for BBC One, scientists recorded sixgill sharks feeding on a dead whale. This meal provides a lot of energy. After eating the whale remains, a sixgill shark can survive a whole year without eating again.

Megamouth sharks spend most of their time in the dark. They have light-producing organs inside their mouths. The light attracts plankton and other small prey.

TIGER SHARKS

Tiger sharks are named for the stripes on their backs. Tiger sharks can grow to be very large. When they are full-grown, they are more than 18 feet (5.5 m) in length.

They also have large appetites. Neil Hammerschlag researches sharks at the University of Miami. He described the feeding habits of tiger sharks: "From sea

Tiger sharks live in warm water. They are most often found near coasts.

birds to sea turtles to dolphins to fish to other sharks, [tiger sharks eat] almost anything in the water."[3] Their knife-like teeth make them dangerous predators. Tiger sharks occasionally attack humans.

Scientists have studied the stomach contents of tiger sharks. They found non-food items inside. These included license plates and tires. Because tiger sharks will try to eat just about anything, they are nicknamed "the garbage cans of the sea."

BONNETHEAD SHARKS

Bonnethead sharks also have unusual diets for sharks. These hammerheads have flat teeth that can crack open shellfish. Bonnethead sharks were also the first shark species known to eat plants. In 2007, researchers studied the diets of these sharks. They found that as much as 62 percent of a bonnethead shark's diet included seagrasses. Samantha Leigh was a graduate student at the time. "This is the first one that we know for sure is [a plant-eater], but it definitely makes me at least want to take a closer look at some other coastal shark species," Leigh said of the group's findings. "It's definitely possible that there are others that could be doing something similar."[4]

SHARK ATTACKS

Some shark species are known to attack humans. Great white sharks, tiger sharks, and bull sharks are considered the Big Three. Together, they make up more than half of all known shark attacks. But overall, shark attacks are extremely rare. Safety practices and public awareness have made attacks even more rare. In 2020, the Florida Museum of Natural History confirmed just 129 attacks worldwide.

3
DEEP-SEA SHARKS

The deep ocean is a difficult place to survive. The water temperature can be very cold. Sunlight becomes very dim at depths below 656 feet (200 m). Below 3,280 feet (1,000 m), there is only darkness. Sharks and other creatures that live in the deep sea must be able to survive these extreme conditions.

These conditions also make the deep ocean difficult to explore. Gavin Naylor is the director of shark research at the Florida Museum of Natural History. "We really do

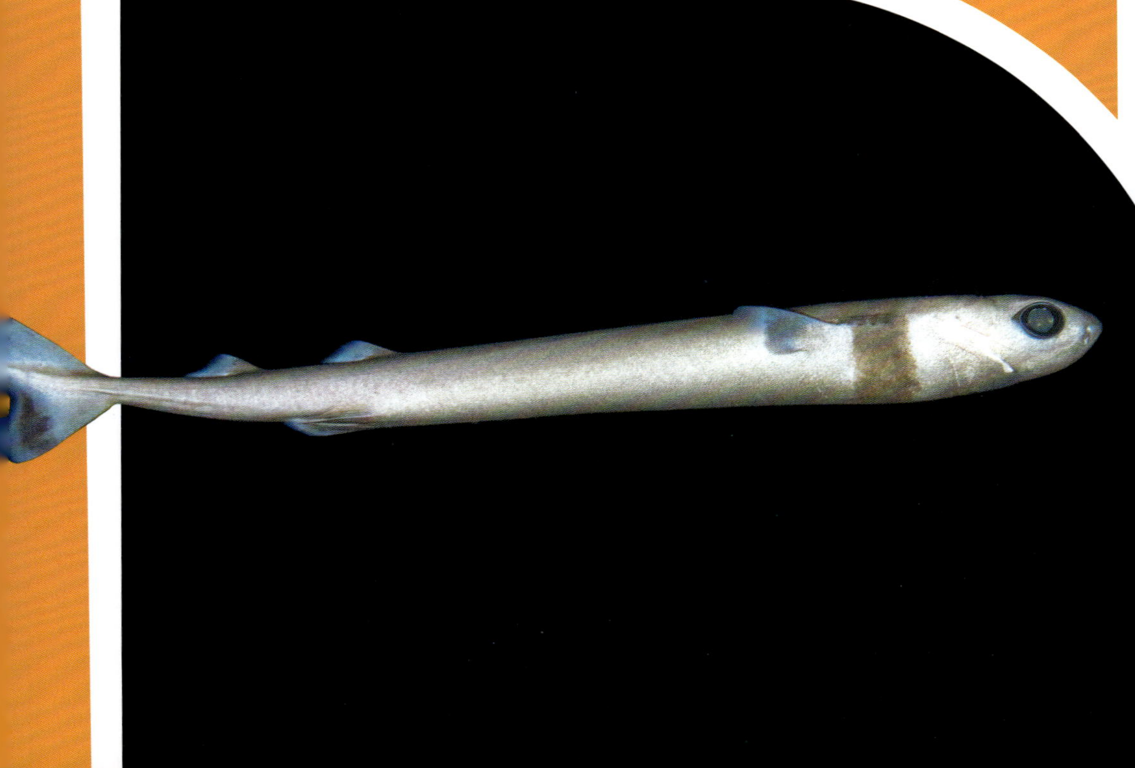

not know much about the ocean—and the little we do know is in the top 500 meters (1,640 feet)," said Naylor. "But when you get down to the deep sea, it's just uncharted. . . . And it is magical."[5]

GREENLAND SHARKS

Greenland sharks live in the cold North Atlantic Ocean. They can survive in water temperatures around 30 degrees Fahrenheit (–1° C). They may dive as deep as 7,200 feet (2,200 m) below the surface.

Greenland sharks have a chemical in their blood that allows them to live in

the cold. The chemical works like antifreeze.
Antifreeze is a substance that lowers the
freezing point of a liquid. The chemical
keeps the shark's blood from freezing. It
also helps these sharks handle the high
pressure of the deep ocean. Their blood
is poisonous. Animals that try to eat a

COOKIE-CUTTER SHARKS

Cookie-cutter sharks live deep underwater. Some
scientists consider these sharks to be parasites.
Cookie-cutter sharks suction their jaws to large
prey. Then, they clamp their triangular teeth
into the creature's flesh. The prey is left with a
circular cookie-shaped wound once the shark has
finished feeding.

Greenland shark can become very sick

and die.

At 24 feet (7 m) long, Greenland sharks

are one of the larger sharks in the ocean.

Many creatures in the deep ocean are

bigger than ones that live in shallow waters. It is easier for big animals to keep their organs warm. They lose less heat to the water.

The cold temperatures affect the lifespan of Greenland sharks. They are the longest-living **vertebrates** on Earth. Scientists think that these sharks can live for more than 500 years. The cold slows down their metabolism, or the rate at which they use energy. Greenland sharks do not have to eat very often.

Julius Nielsen is an ocean scientist at the University of Copenhagen in Denmark.

His team of researchers discovered a Greenland shark thought to be 400 years old. "We had our expectations that we were dealing with an unusual animal," Nielsen said of their findings. "But I think everyone doing this research was very surprised to learn the sharks were as old as they were."[6]

Greenland sharks are often blind. Parasites eat their eyes. Since the deep ocean is so dark, Greenland sharks do not rely on their eyesight to hunt. Some scientists think that these eye parasites help Greenland sharks in other ways. Small animals may approach the sharks to eat

Frilled sharks usually swallow their prey whole.

the parasites. Greenland sharks are able to

suck in the prey as they get close.

FRILLED SHARKS

Scientists sometimes refer to frilled sharks

as living fossils. These sharks are thought

to have first lived 80 million years ago.

They have long, eel-like bodies and small

fins. Most shark species have five gill slits.

Frilled sharks have six pairs of gills. This

helps them take in more oxygen from their

surroundings. There is less oxygen in their

deep ocean habitat.

Frilled sharks have unusual teeth. Each

tooth has three points. This gives the sharks

a terrifying appearance. Simon Boag works

in the fishing industry. He described a

frilled shark in 2015. "It has 300 teeth over

twenty-five rows, so once you're in that

mouth, you're not coming out."[7]

Pygmy sharks glow on their undersides.

GLOWING SHARKS

Pygmy sharks are some of the smallest sharks in the ocean. They grow to a maximum length of just over 10 inches (25 cm). They can survive at depths of 32,600 feet (9,940 m) below the surface.

Jasmin Graham is the president of Minorities in Shark Sciences. She says, "I always share with people that 75 percent of sharks are less than 3 feet [0.9 m] long. . . . The vast majority of sharks . . . aren't very big, and most live in the deep sea where you're never going to encounter them."[8]

Pygmy sharks have organs that glow. The light attracts prey. It also helps pygmy sharks hide from predators. When viewed from below, the sharks blend in with the sunlight from above.

The deep ocean is home to other sharks that appear to glow. The chain catshark is one of more than 100 catshark species. It absorbs blue light. It uses this light to emit a green glow that helps it attract mates. The viper dogfish also has organs that glow. In 2020, researchers discovered that kitefin sharks produce their own light. Kitefin sharks are the largest vertebrates known to have this ability.

BIOLUMINESCENT OR BIOFLUORESCENT?

Pygmy sharks are bioluminescent. This means they produce their own light. Other sharks are biofluorescent. They do not make their own light. Instead, they reflect sunlight to glow. Biofluorescent sharks include chain catsharks and swell sharks.

REEF DWELLERS

Sharks that live close to the shore have unusual adaptations to help them survive in coral reefs. Some use camouflage to surprise their prey. Some lie still on the ocean floor. Others use their fins to move along reef bottoms.

WOBBEGONGS

There are twelve species of wobbegongs.

They are also known as carpet sharks.

Wobbegongs have flat bodies that help

them hide on the floors of coral reefs.

Their coloration also helps them blend in

with sand and rocks. Wobbegongs have

The scientific name of wobbegongs is Orectolobus maculatus. The name means "stretched out spot."

barbels. These whisker-like extensions surround the wobbegong's head.

Wobbegongs are mostly found in the Indian Ocean near Australia. The word *wobbegong* is thought to have roots in an Aboriginal language. Aboriginal peoples are the original peoples of Australia. *Wobbegong* means "shaggy beard."

The wobbegong's weird appearance is what makes the shark a dangerous hunter. It lies in wait in rocky crevices near the ocean floor. It attracts curious prey with its barbels. The wobbegong also lures prey with its tail. While hiding in the rocks, it

slowly moves its tail back and forth. The

movement tricks small fish. They think there

is another fish in the rocks. The fish think

the rocky area is safe, so they swim closer.

The wobbegong opens its mouth and

sucks up the prey once it is close by.

ANGEL SHARKS

Angel sharks are another type of ambush predator. They surprise their prey by hiding on the ocean floor. Like wobbegongs, angel sharks have flat bodies. They look similar to rays.

Angel sharks bury themselves in the sand. Their gray, brown, and black colors blend into the ocean floor. Their eyes are located on the top of their bodies. They can see while the rest of their bodies are buried.

Some sharks need to keep swimming to breathe. They need water to constantly flow over their gills. Angel sharks are able to breathe while staying still. They have special muscles that pump water over their gills.

They also breathe through structures called spiracles that are on the top of their heads. These adaptations allow angel sharks to lay still while they wait for their prey.

Angel sharks mostly eat small fish. They wait for the fish to swim close. Then, they

RAY OR SHARK?

Sharks with flat bodies may look like rays. But they have two major differences. The **pectoral fins** of rays are attached to their heads. In sharks, these fins are attached to their bodies. Their gills are also located in different places. Rays have gills on the undersides of their bodies. Shark gills are on their sides.

burst toward the fish. They swallow their prey whole in a single gulp.

WALKING SHARKS

Scientists have observed some sharks using their pectoral fins to crawl on the ocean floor. Horn sharks are not strong swimmers. They use their fins to move across rocks. Nurse sharks sometimes walk on their fins as well. Their fins help position them before they attack their prey.

But the epaulette shark is most famous for its walking abilities. It is sometimes called the walking shark. It spends most

of its time in shallow water. The epaulette

shark can crawl across land on its fins for a

short time. It can move between separated

tide pools.

These sharks are not very large.

They typically do not grow longer than

35 inches (90 cm). But being able to survive in shallow water is a huge advantage. Tide pools have low oxygen levels. Most ocean animals cannot survive in these conditions. Epaulette sharks can stay in waters with very low oxygen levels for more than three hours. Predators cannot reach them there.

Scientists discovered four new species of walking sharks between 2008 and 2020. Shark scientist Gavin Naylor is excited about these findings. He said, "[The shallow waters] may be the one place in the world where [the development of new species] is still going on for sharks."[9]

Epaulette sharks are usually a cream or brown color. They have a large black spot on each side that looks like an eye to frighten predators.

But there is still much to explore in the deep sea. From the shallows to the deep, the ocean is home to many strange and unusual sharks. Scientists are excited to discover more about the weird sharks in Earth's oceans.

GLOSSARY

adaptations

parts of an animal or a plant that allow it to survive well in its habitat

migrate

to move from one region to another to find food or to breed

organs

body parts that work to perform a certain function

parasites

animals or plants that live in or on and take energy from another animal or plant

pectoral fins

a pair of fins located near the head that help a fish swim and turn

predators

animals that kill and eat other animals

prey

an animal that is killed and eaten by other animals

species

a group of animals of the same kind

vertebrates

animals with a backbone

SOURCE NOTES

CHAPTER ONE: SHARK HEADS AND SHARK TAILS

1. Quoted in Cara Giamino, "The Pros and Cons of Swimming with a Hammerhead," *New York Times*, September 21, 2020. www.nytimes.com.

CHAPTER TWO: FILTER FEEDERS AND WEIRD EATERS

2. Quoted in Hazel Pfeifer, "How NASA Technology Can Help Save Whale Sharks—The World's Largest Fish," *CNN*, February 9, 2021. www.cnn.com.

3. Quoted in Nikk Ogasa, "While Some Sharks Flee, Tiger Sharks Brave Stormy Seas," *Scientist*, May 21, 2021. www.the-scientist.com.

4. Quoted in Veronique Greenwood, "The Omnivorous Sharks That Eat Grass," *New York Times*, September 6, 2018. www.nytimes.com.

CHAPTER THREE: DEEP-SEA SHARKS

5. Quoted in Haley Cohen Gilliland, "Scientists Tag Deep-Sea Shark Hundreds of Feet Underwater—A First," *National Geographic*, September 5, 2019. www.nationalgeographic.com.

6. Quoted in Rebecca Morelle, "400-Year-Old Greenland Shark 'Longest-Living Vertebrate,'" *BBC*, August 12, 2016. www.bbc.com.

7. Quoted in Bill Chappell, "Rare and 'Horrific': Frilled Shark Startles Fishermen in Australia," *NPR*, January 21, 2015. www.npr.org.

CHAPTER FOUR: REEF DWELLERS

8. Quoted in Kara Norton, "Meet the Women Diversifying Shark Science," *NOVA*, July 30, 2021. www.pbs.org.

FOR FURTHER RESEARCH

BOOKS

Yvette LaPierre, *Prehistoric Sharks*. San Diego, CA: BrightPoint Press, 2023.

Madeline Nixon, *Hammerhead Shark*. New York: AV2, 2019.

Shark. New York: DK Publishing, 2022.

INTERNET SOURCES

"10 Weirdest Sharks in the World – And Top 5 Weirdest Extinct Sharks," *Sharkwater*, 2021. www.sharkwater.com.

Brianna Elliott, "Photos: Introducing Deep-Sea Sharks, Some of the Wildest Looking Fish in the Oceans," *Oceana*, August 13, 2014. https://usa.oceana.org.

Andrea Silen, "Goblin Shark," *National Geographic Kids*, 2015. https://kids.nationalgeographic.com.

WEBSITES

American Shark Conservancy
www.americansharkconservancy.org

American Shark Conservancy promotes awareness of sharks and rays. It works to improve ocean habitats for all creatures.

Ocean Conservancy
https://oceanconservancy.org

Ocean Conservancy is an organization dedicated to improving the health of the ocean for sharks and other creatures.

Oceana
https://oceana.org

Oceana strives to protect the world's oceans. It works with scientists to create policies that help the ocean and its resources.

INDEX

IMAGE CREDITS

Cover: © Animal Stock/Alamy
5: © Martin Prochazkacz/Shutterstock Images
7: © Nicolas Voisin 44/Shutterstock Images
9: © Rich Carey/Shutterstock Images
11: © Alex Vog/Shutterstock Images
13: © Doug Perrine/Blue Planet Archive
14: © Martin Voeller/iStockphoto
17: © K3 Star/Shutterstock Images
19: © Marko Steffensen/Blue Planet Archive
25: © Rich Carey/Shutterstock Images
26: © Colin Marshall/FLPA/Science Source
28: © Martin Prochazkacz/Shutterstock Images
33: © Kaschibo/Shutterstock Images
37: © Jeff Milisen/Blue Planet Archive
40: © Saul Gonor/Blue Planet Archive
43: © Rohit Kushwaha/iStockphoto
45: © Doug Perrine/Blue Planet Archive
49: © Serge UW Photo/Shutterstock Images
51: © Stewart Kirk/Shutterstock Images
55: © Richard Whitcombe/Shutterstock Images
57: © Slow Motion Gli/Shutterstock Images

Chelsea Xie lives in San Diego, California. She enjoys reading books on the beach, far away from sharks.